AF307772

Architectures

Gabrielle von Bernstorff

Printed and Published by

BoD – Books on Demand,

Norderstedt

ISBN: 9783759751911

2024

Further books by Gabrielle von Bernstorff

Hope for Moria: Städte anstelle von Barbarei

Hamburgs Hafencity Jubiläumsausgabe

www.gabrielle-von-bernstorff.weebly.com

Architectures

Thanks-Givings

Architecture does not happen in a vacuum. Hundreds if not thousands of people participate in its making. Thanks to all those who have participated in the projects shown here, friends, colleagues, partners, administrators, officials, clients, consultants, workers and builders. These works would not exist without you.

Introduction

Architecture is a public art. Some buildings make us happy, others make us sad. We live architecture almost the way we breathe, it surrounds us endlessly. In buildings and with buildings we dream, sleep and wake. We eat drink, celebrate, love, dispute and weep. Children play and do their homework. Families do their thing, artists do their thing, workers do their thing. Hardly anything we do is not connected to architecture, from the simplest hut to the most lavish concert hall. There is architecture that makes our spirit soar, other buildings make us reflect and contemplate. And all this although architecture somewhere always steps into the background because living is at the forefront of it all.

Each building is an architecture onto its own. Every beginning is a fresh start, guided by light, site, nature, material and its, so called, use. Use is a euphemism for living. Even in the most humble of shacks,- we see how someone lives. The care taken in special arrangements, the care taken in the making of a place is always the presence of someone alive, someone living here.

A building is a social sculpture in the Beuysian sense, a fulcrum and hinge with which we engage with the world and the universe. The making of a building is a long and intense process with many participants, from the clients to the architects, consultants and builders. Architecture is social process, not only in its making.

Architecture is poesis and techne, it is weaving and poetry in the making. 'A dome is not made by building it', Louis Kahn.

So, there is something other about architecture, that at its best touches our hearts and souls. I hope that you will be touched.

Switzerland 2024

Gabrielle von Bernstorff

A Museum

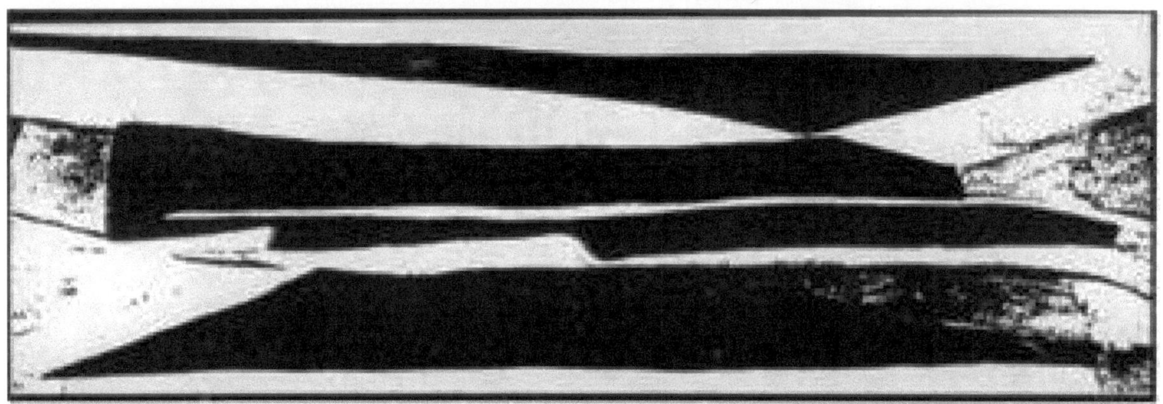

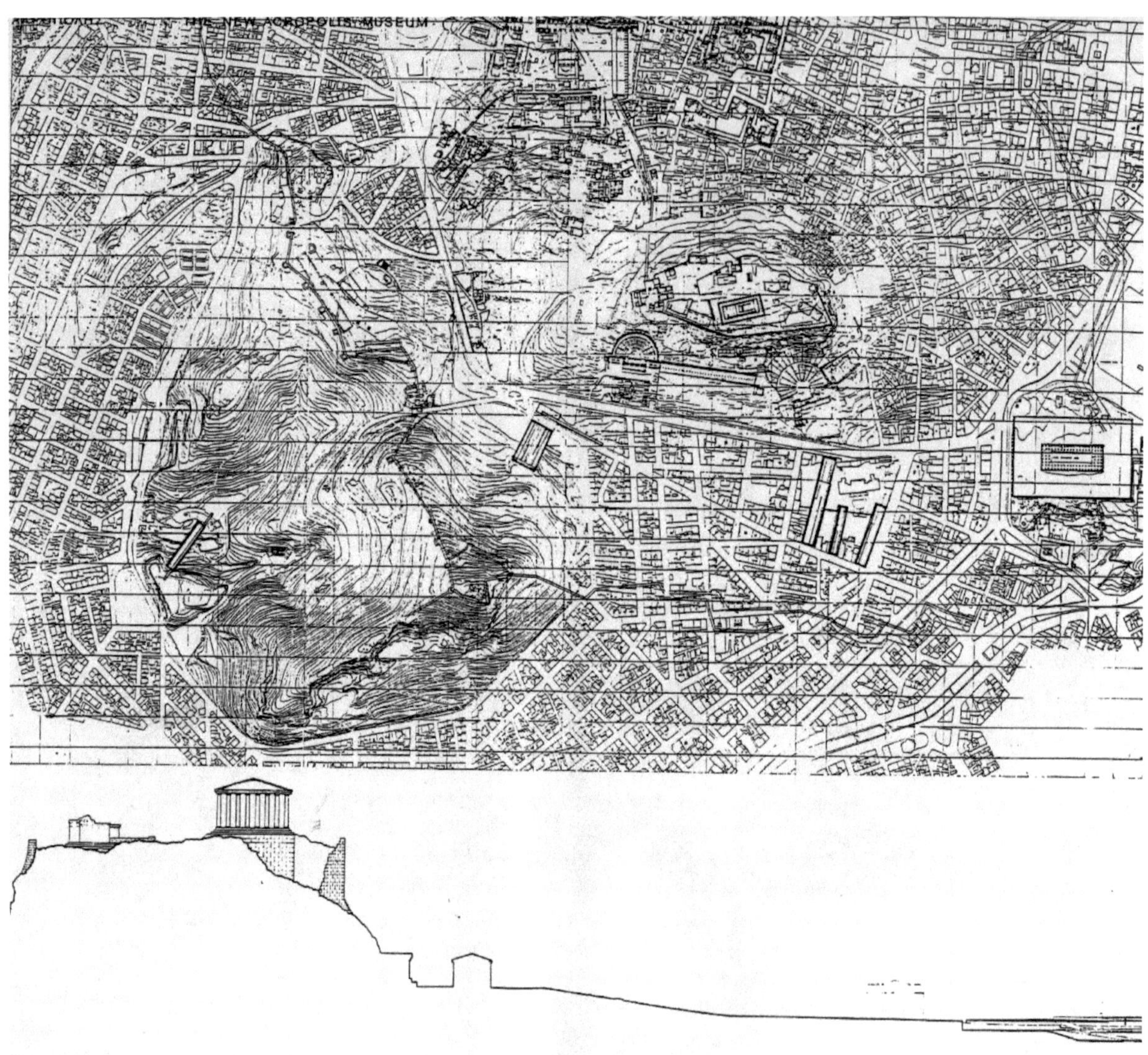

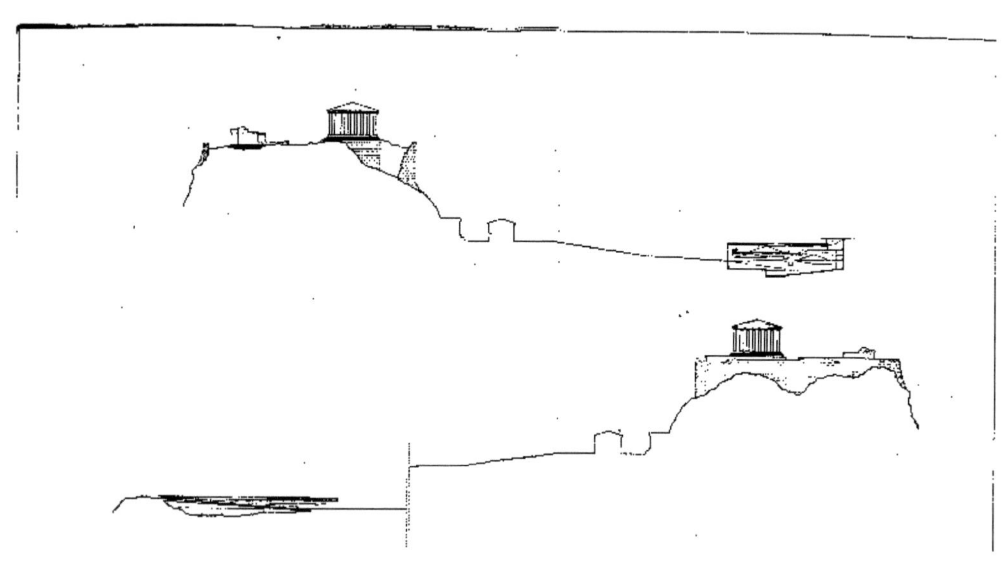

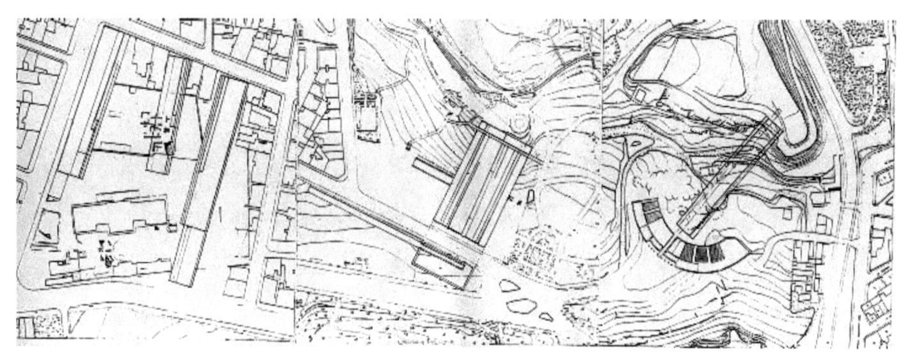

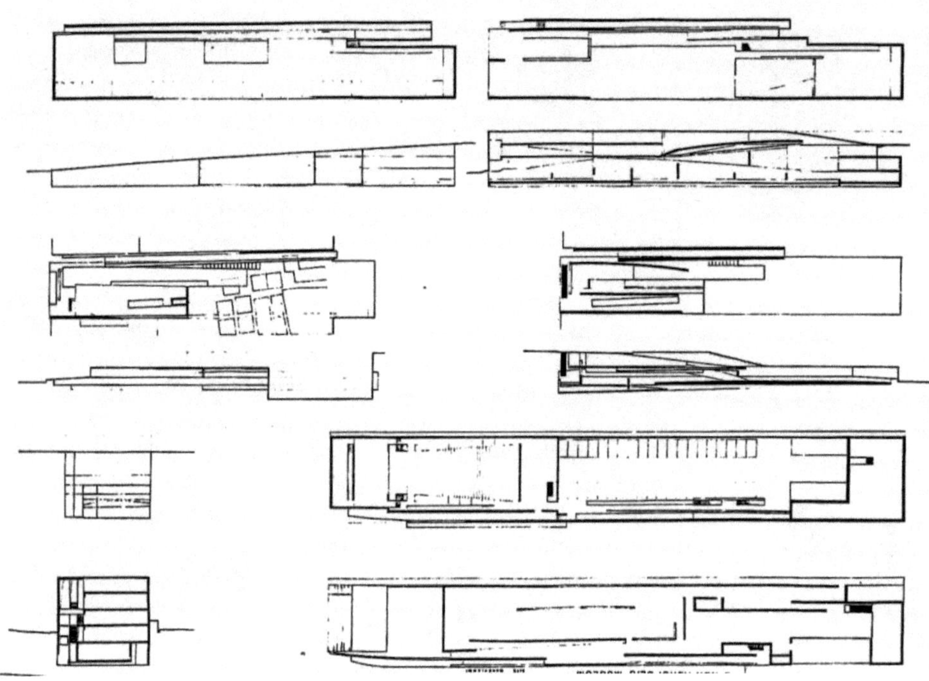

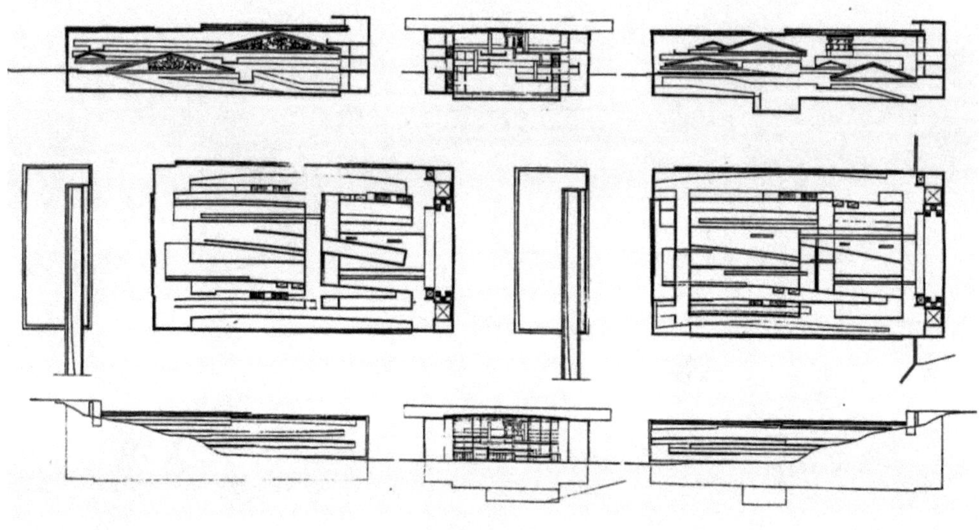

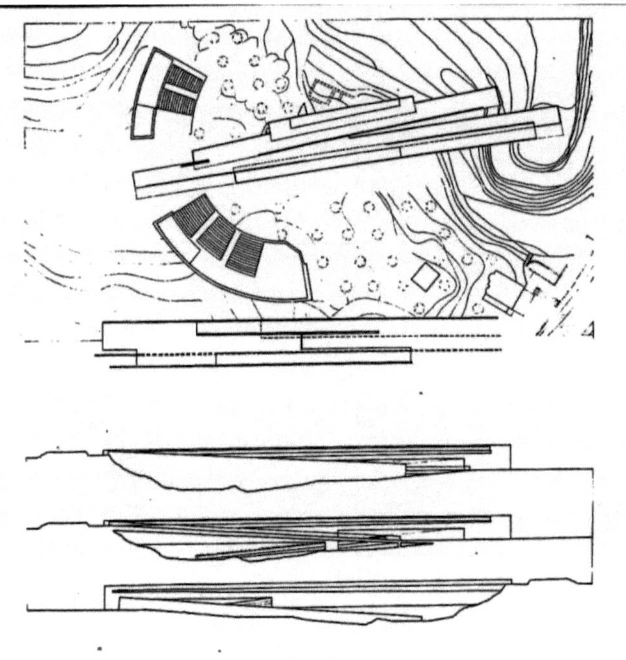

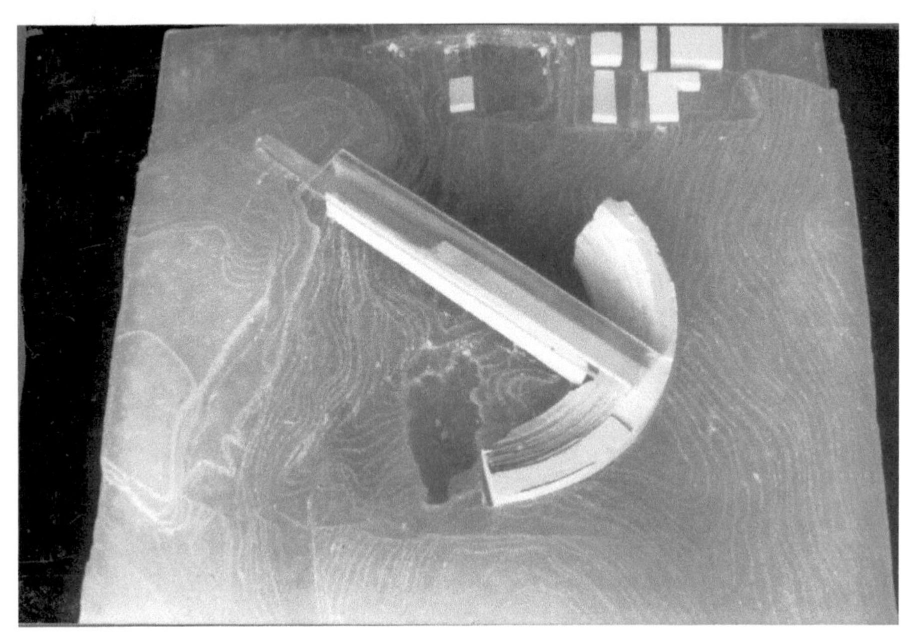

A House

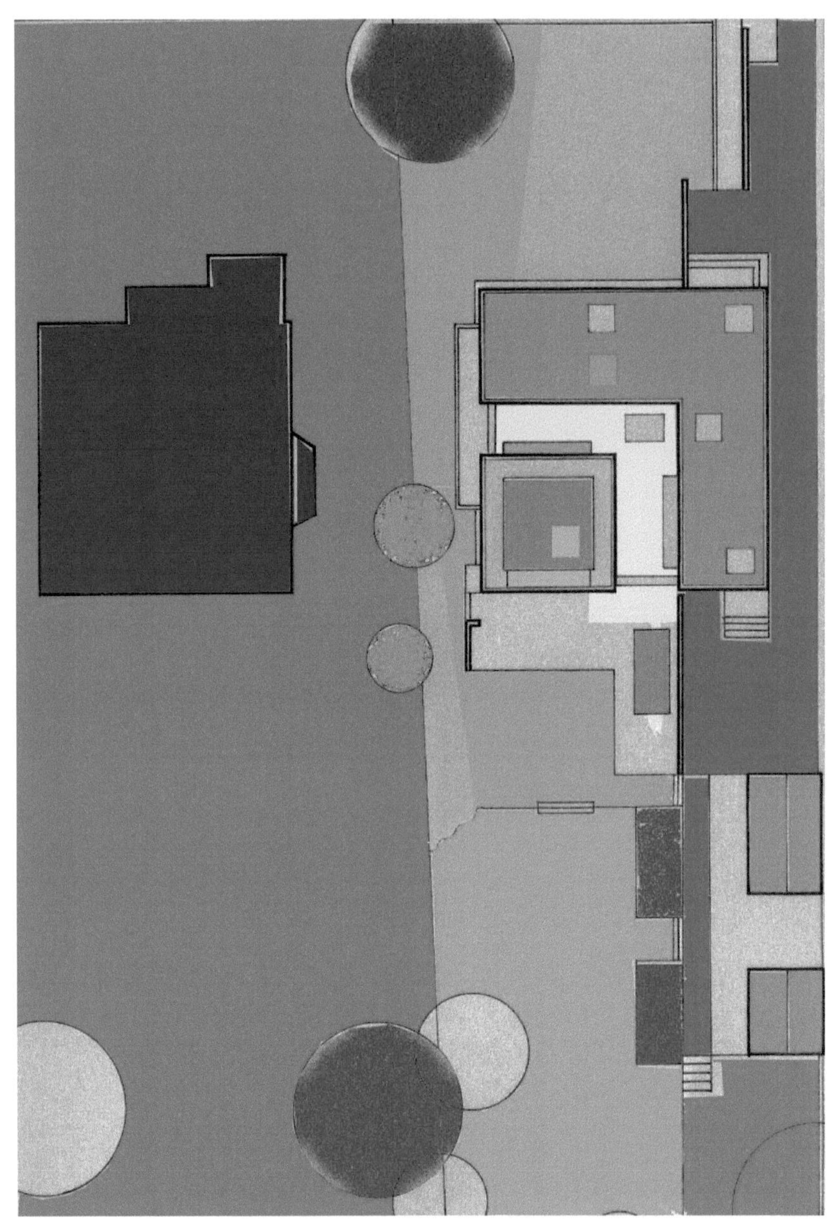

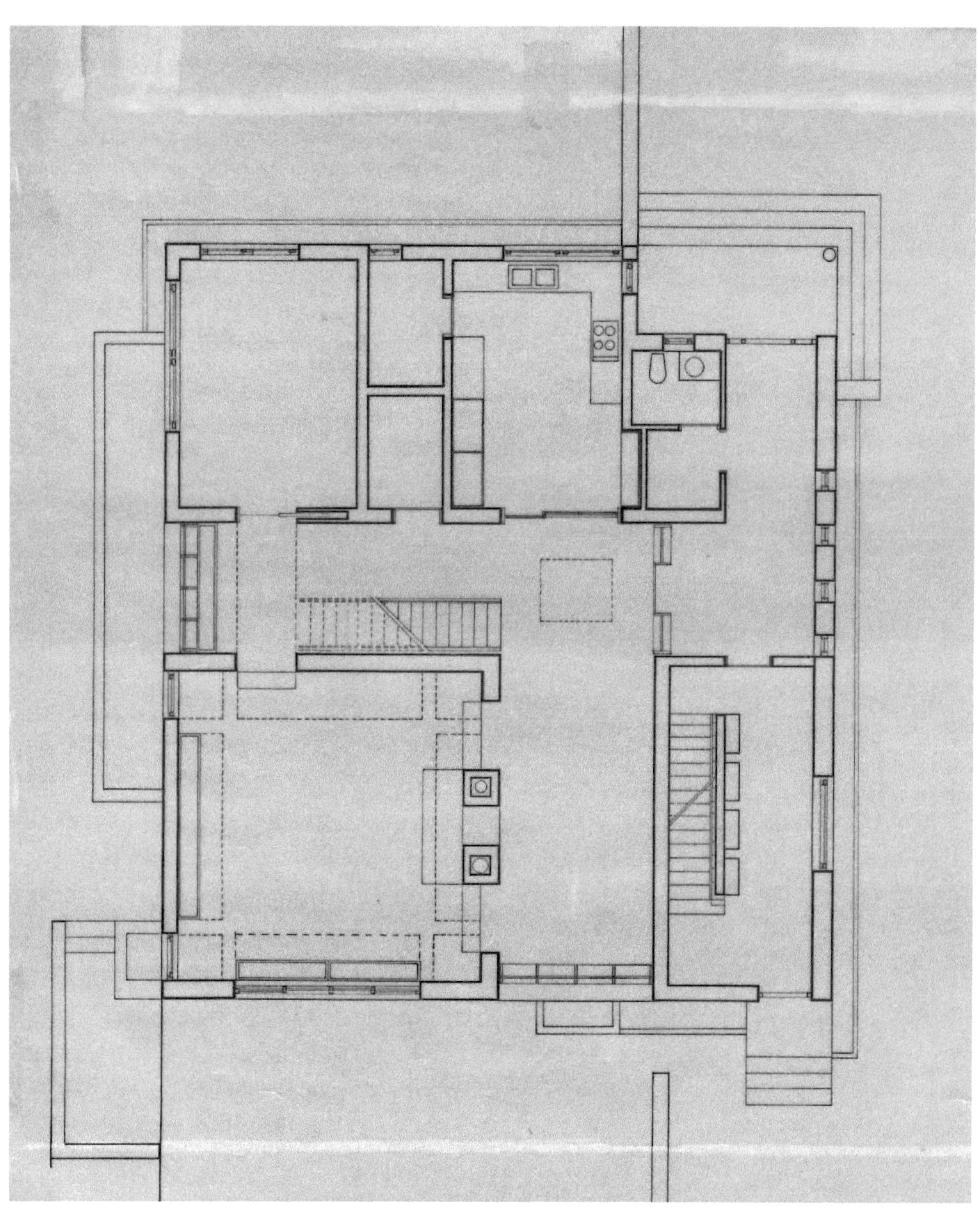

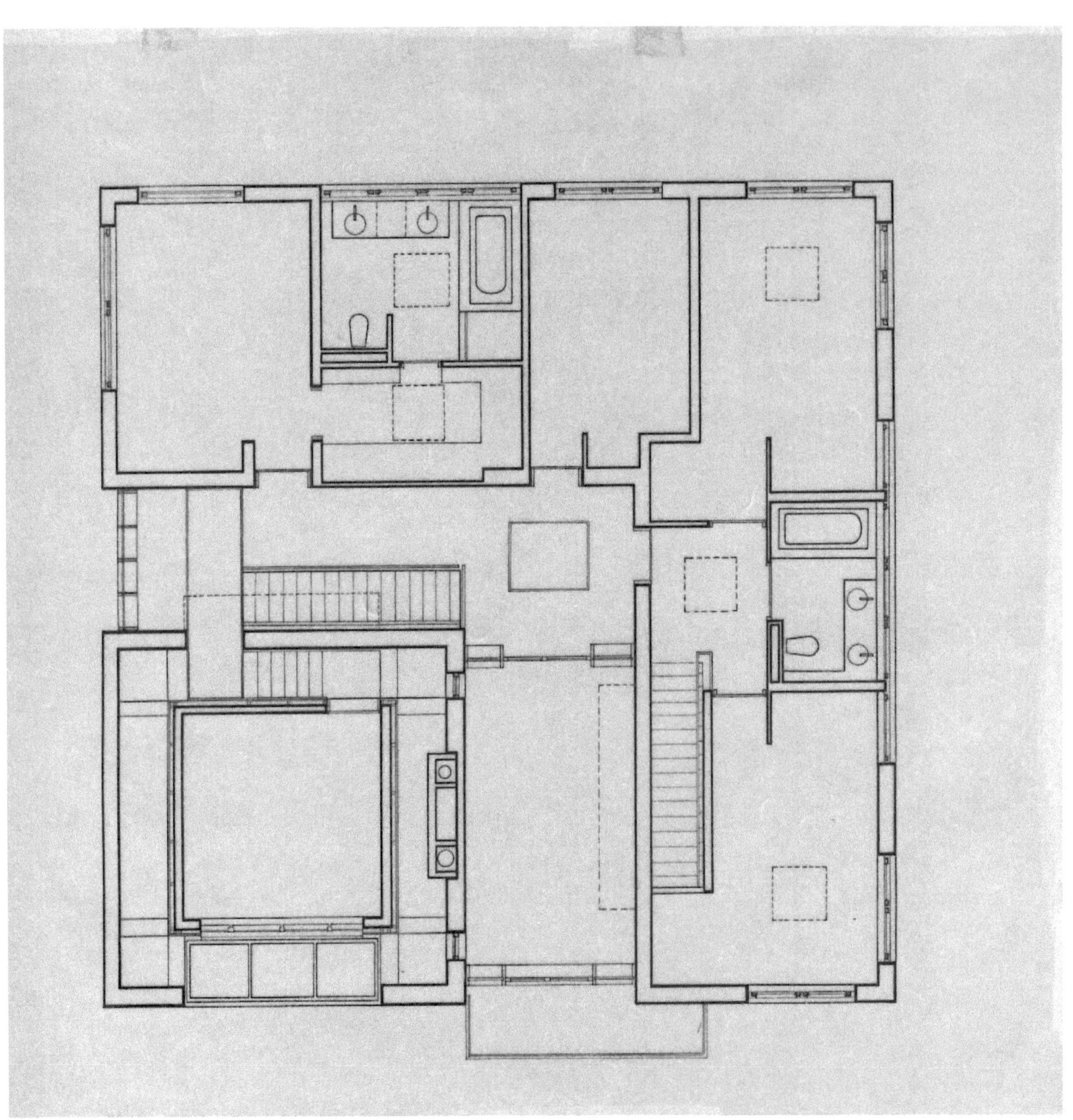

A Cultural Centre

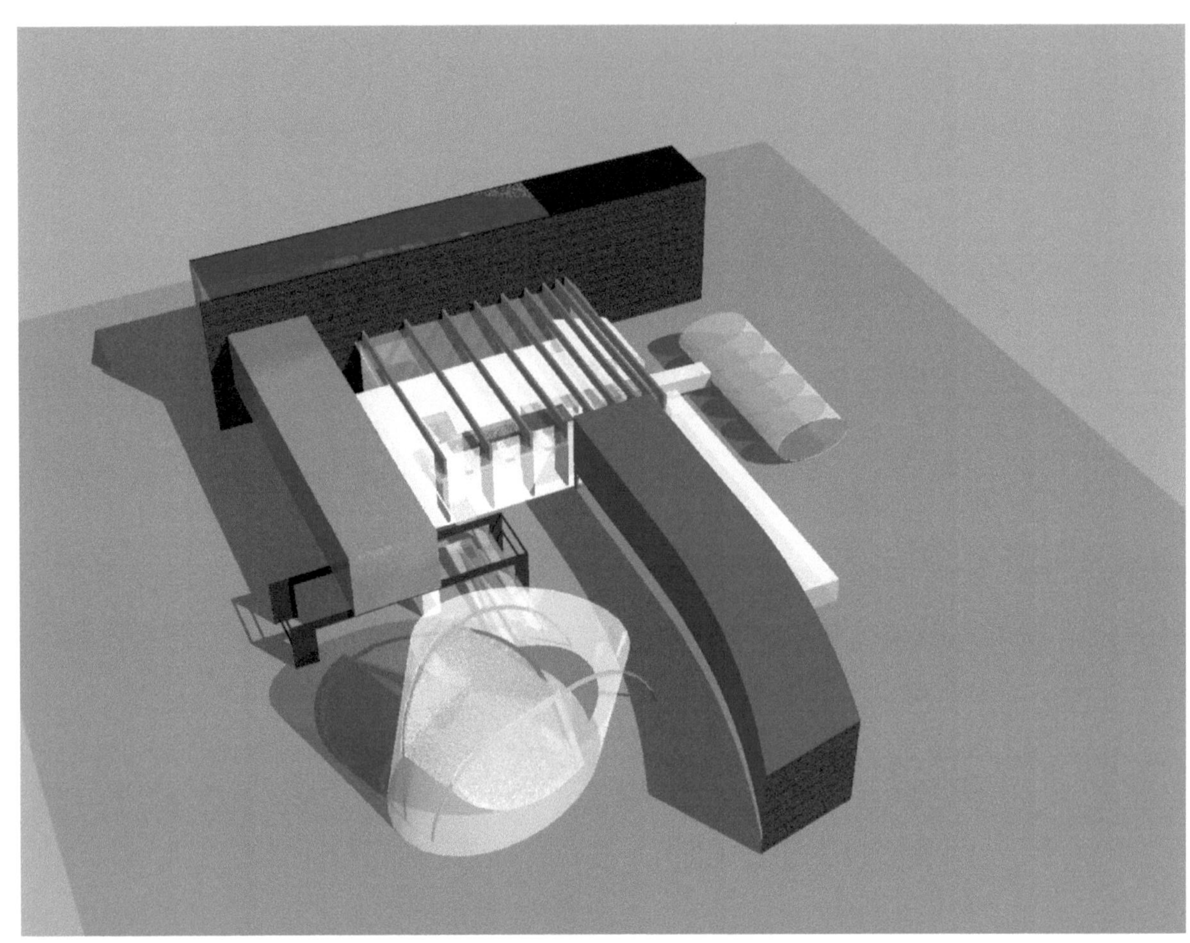

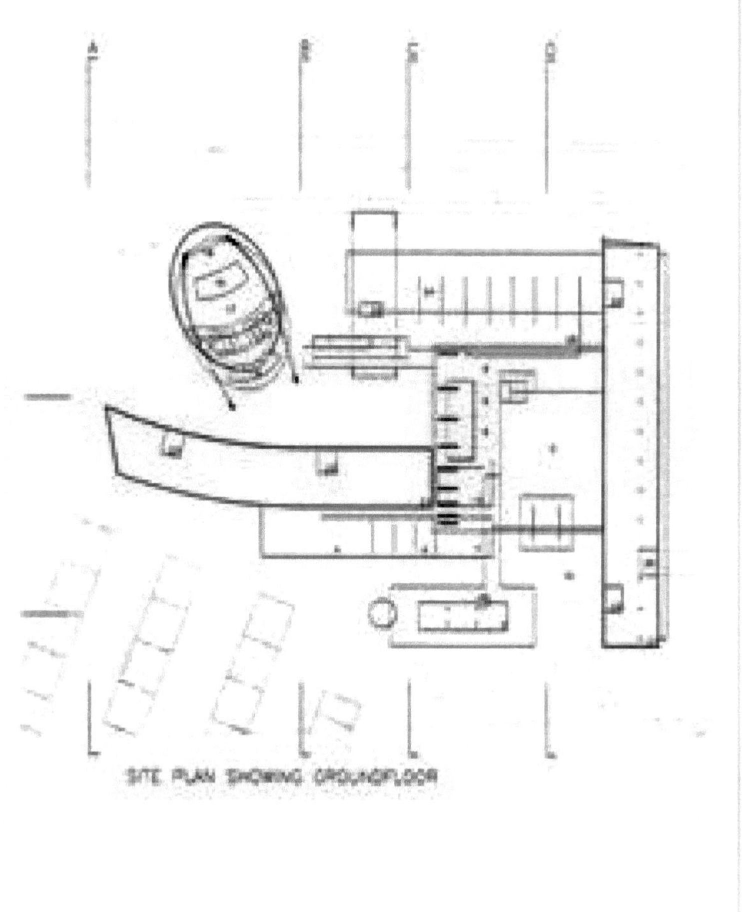

SITE PLAN SHOWING GROUNDFLOOR

A Building

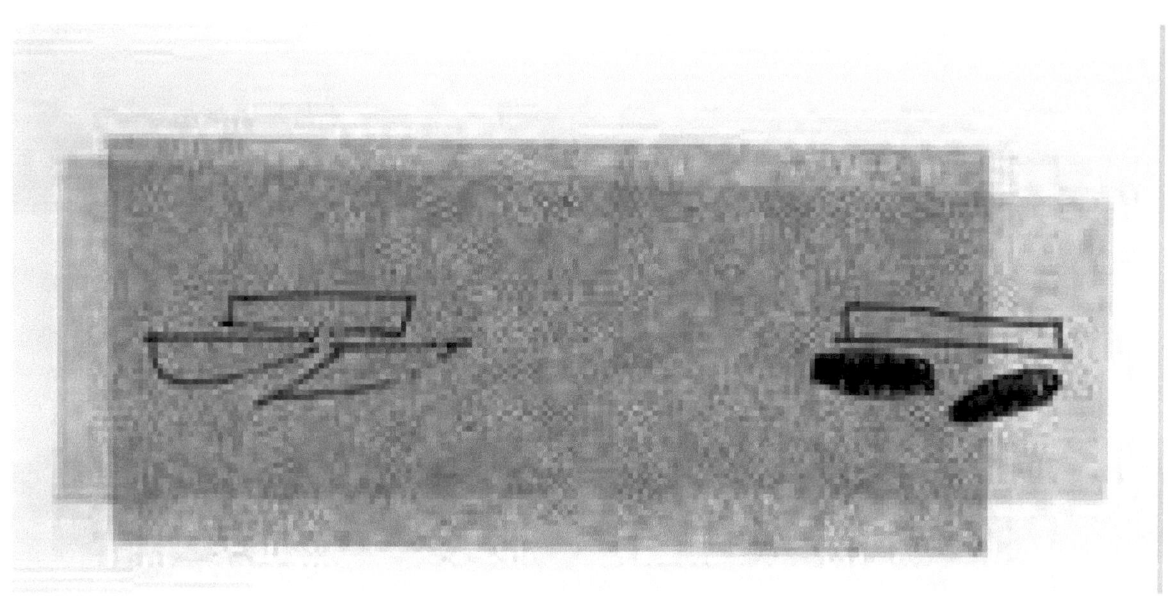

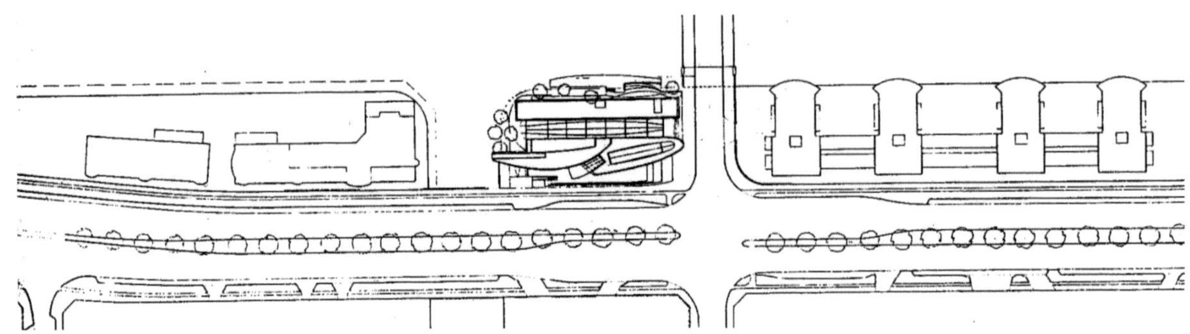

An Artist Residence

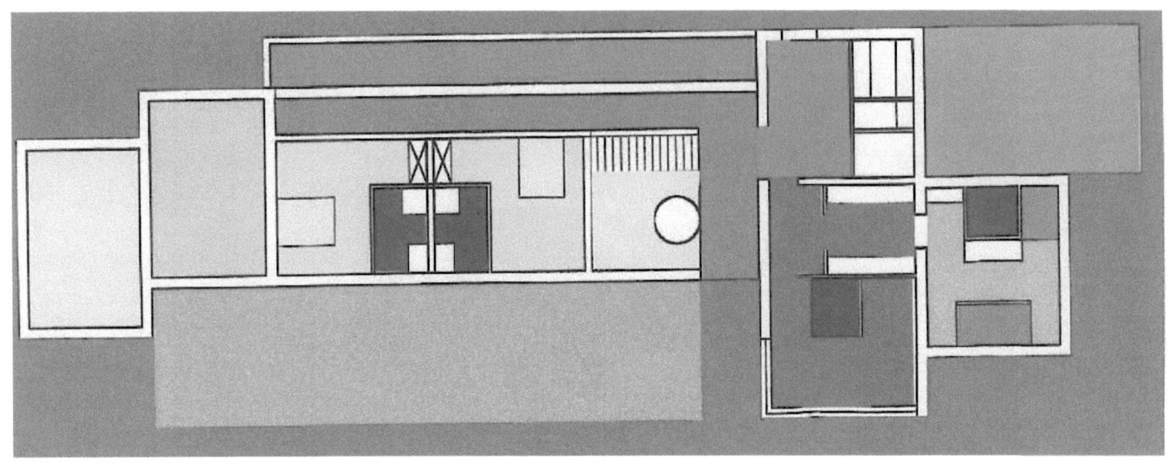

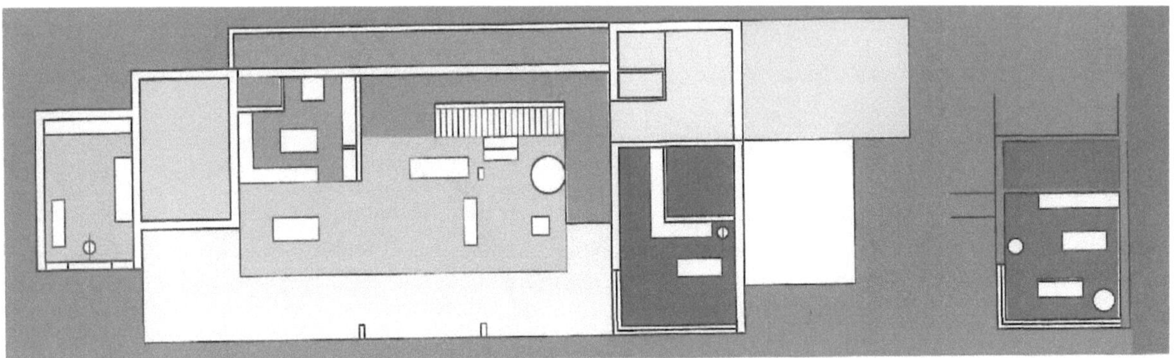

A Theatre

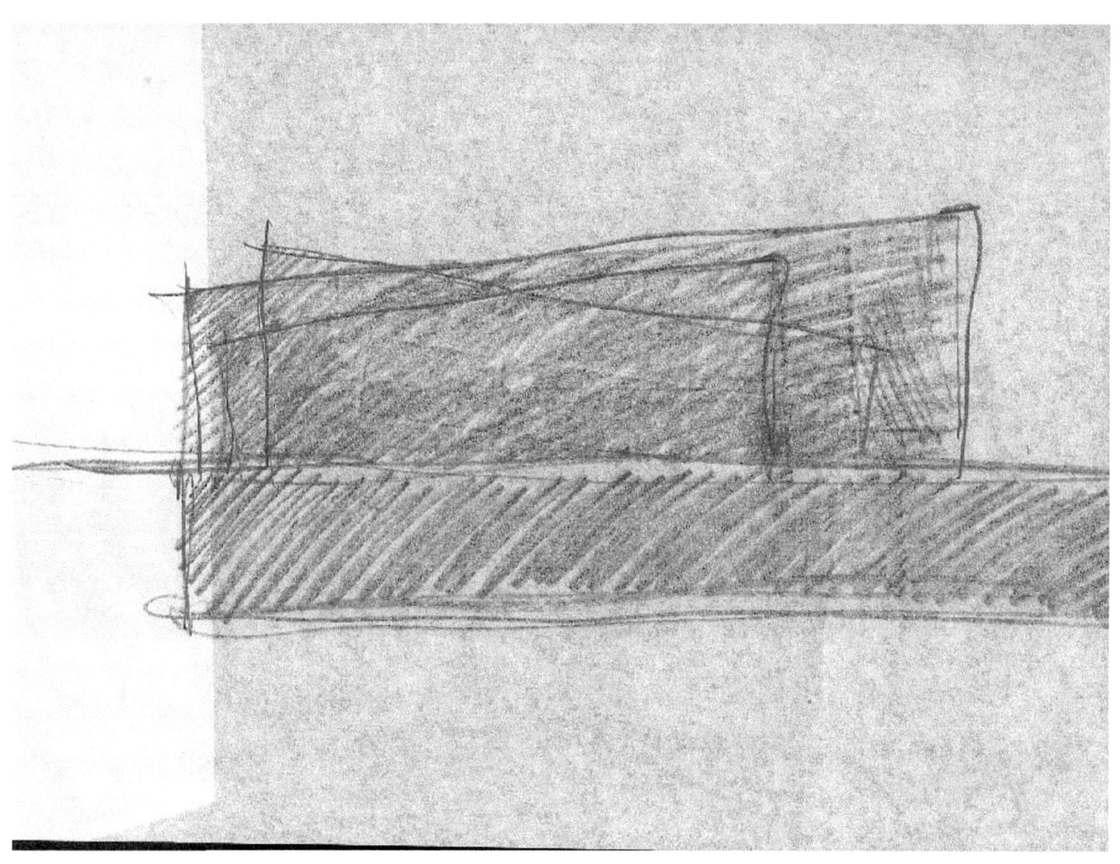

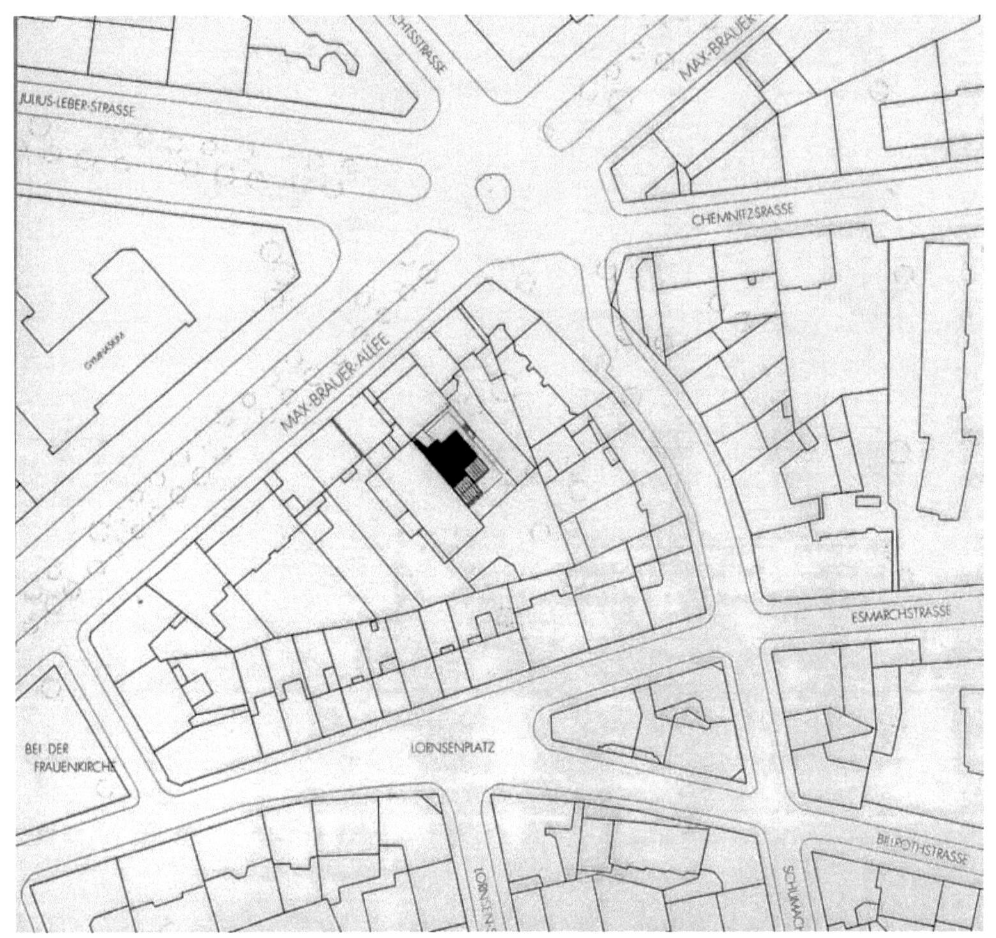
JULIUS-LEBER-STRASSE
MAX-BRAUER
CHEMNITZSTRASSE
GYMNASIUM
MAX-BRAUER-ALLEE
ESMARCHSTRASSE
BEI DER
FRAUENKIRCHE
LORNSENPLATZ
LORNSENS
BELLROTHSTRASSE
SCHUMACH

Housing

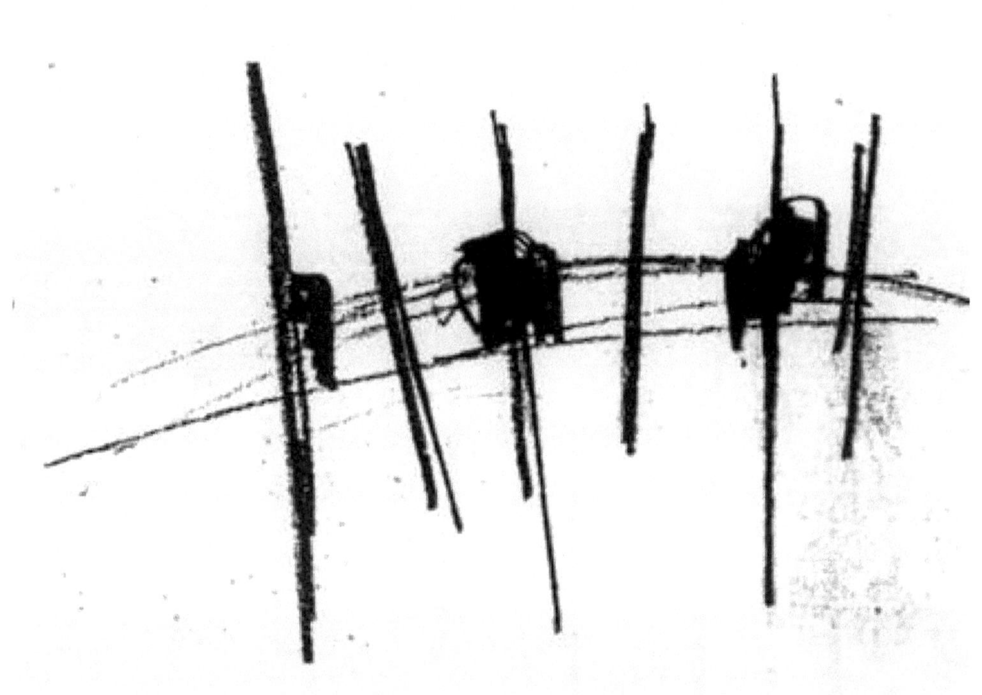

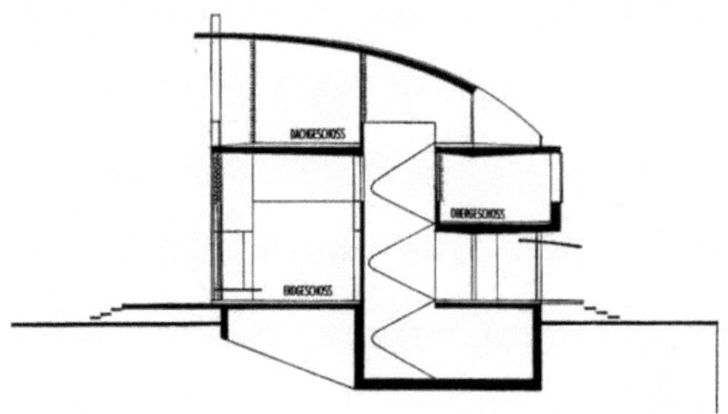

DACHGESCHOSS
OBERGESCHOSS
ERDGESCHOSS

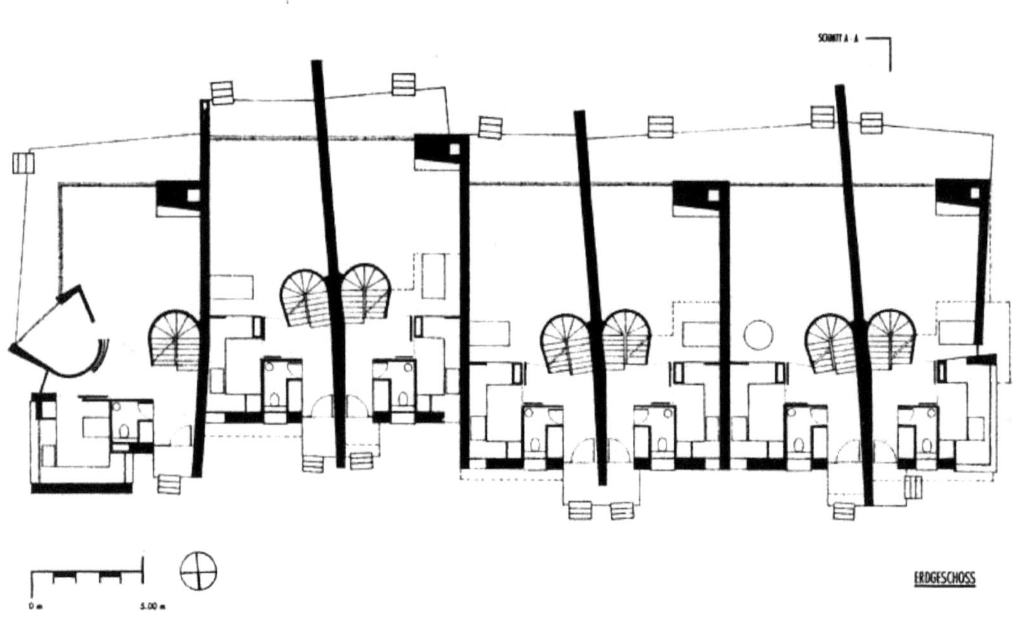

SCHNITT A-A
0 m
5.00 m
ERDGESCHOSS

Two Generational Houses

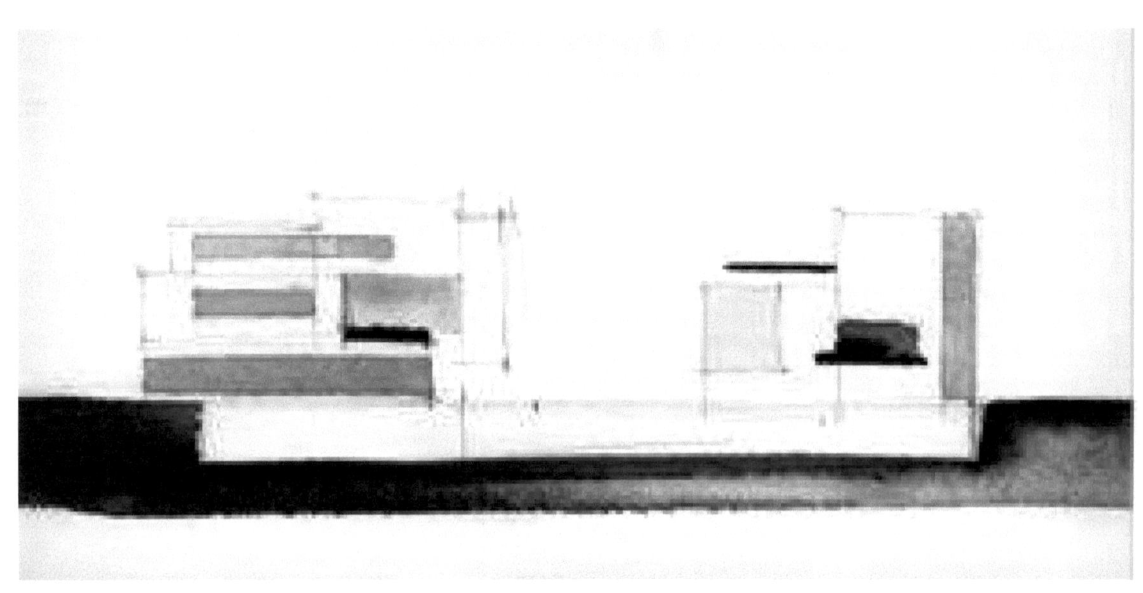

A Work-Home

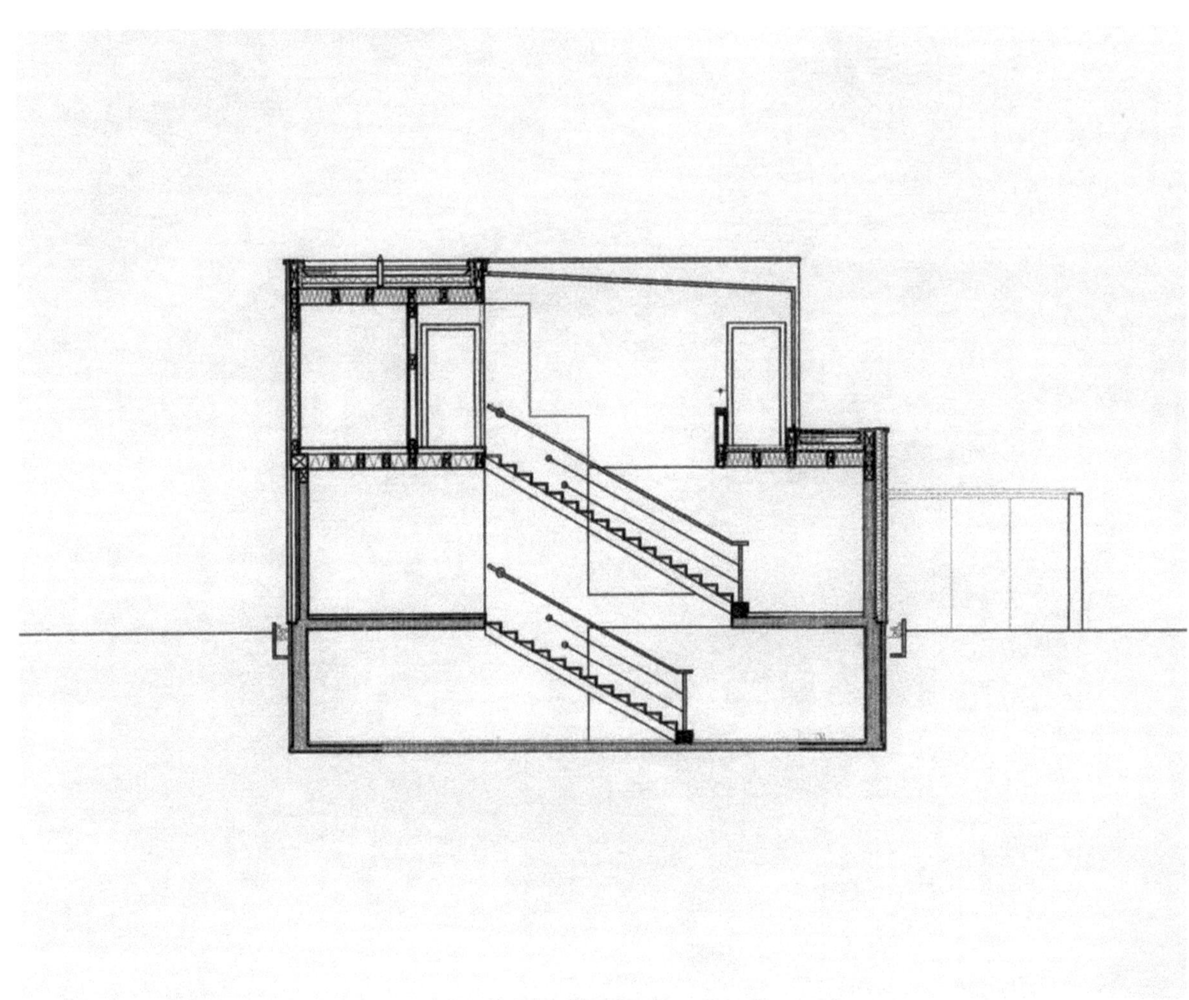

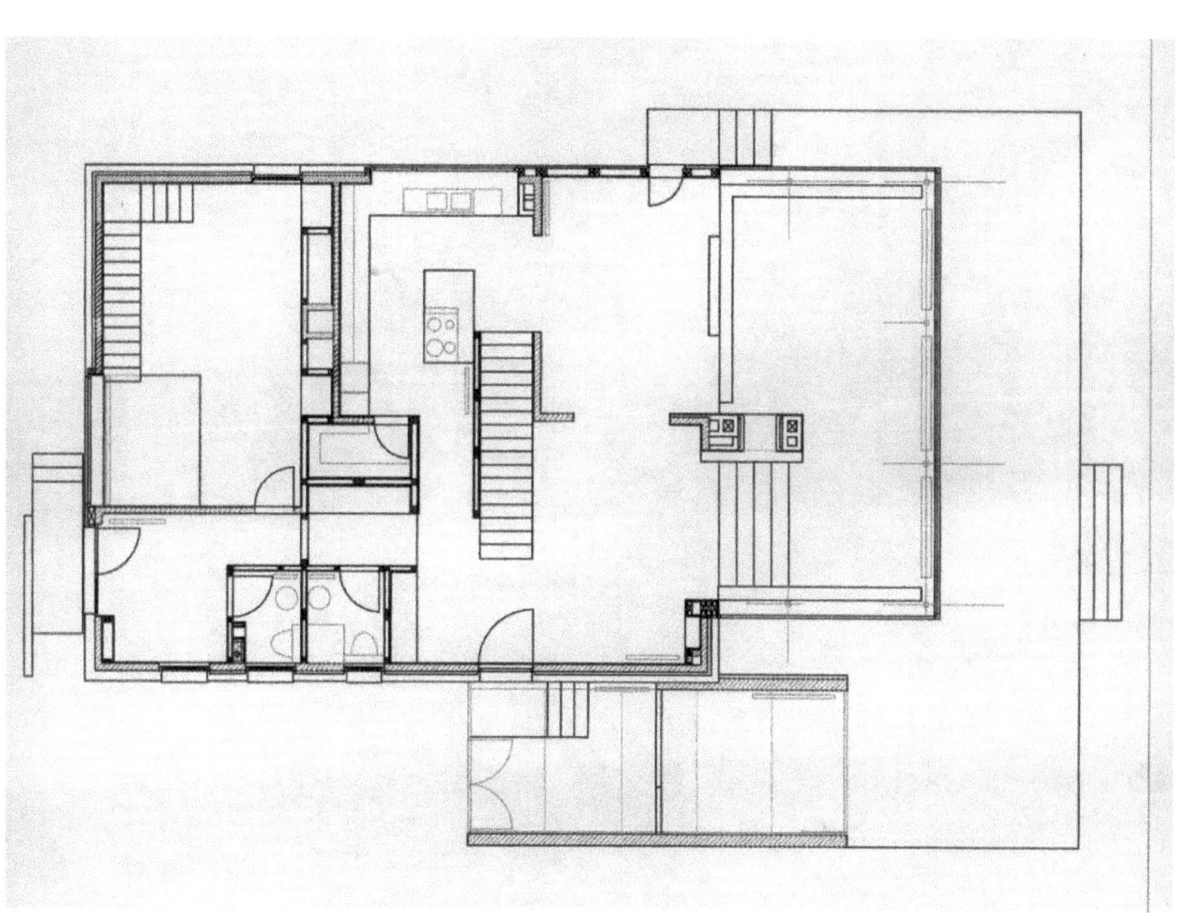

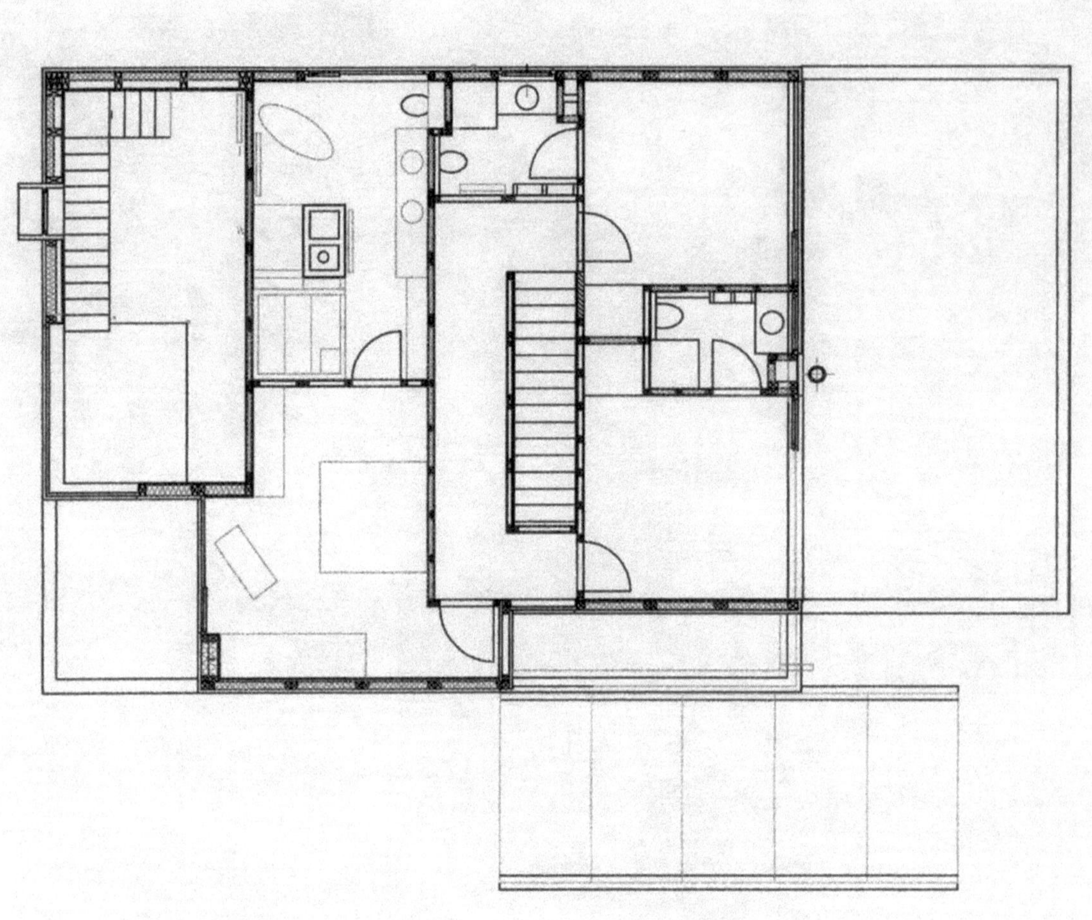

Postscript

Every building is a fresh start, is something unique and a chance to explore, create, be inspired, imagine and dream. According to Gaston Bachelard we must invent that which we wish to find.

Architecture, even if modular, is not like car fabrication. Something about it is hand-made. That is what distinguishes architecture from the digital universe. Architecture is connected to the elements, to nature and to human beings in the flesh.

Architecture usually is made differently from regular commercial goods. It is a journey from an idea to a finished 'product'. I love and cherish sketches, hand made models and drawings that do not imitate the 'real'. Architecture is a process from the immaterial to the material and our life with buildings is, at least in part ideational. That is why the drawings and photos of buildings are all in black and white here. The idea is to be removed somewhat from our enslavement to the real, which does not ignore that architecture is very real!

Shown here is a cross section of some of the projects I have had the privilege to be a part of, experiences I am deeply grateful for.

For more go to my website www.gabrielle-von-bernstorff.weebly.com